TREASURE HUNT

FINDING LOVE GOD'S WAY

https://thetreasurehunt.life/
Print Book Edition 2025
ISBN: 979-8-9987543-7-1

Printed in the United States Of America
Published by Living Water Book Publishing LLC
Livingwaterbooks.org

Religion Christian Ministry and Discipleship
Personal Growth/Men
Family and Relationships General

Please note that Living Water Books capitalizes specific pronouns in scripture that refer to the Father, Son, and Holy Spirit and may differ from some publishers' styles.

TREASURE HUNT

FINDING LOVE GOD'S WAY

JAMIN NEWSOME

TABLE OF

CONTENTS

> CHASING WOMEN, MONEY, AND STATUS WILL NEVER SATISFY THE DEEP HUNGER IN **HIS SOUL**
>
> **JAMIN NEWSOME**

PRELUDE 00

This book is for the man who's tired of the games, the heartbreak, the shallow connections—and wants something real.

Something sacred.

A relationship built not on lust, but on love. Not on impulse, but on purpose. Not on charm, but on Christ.

INTRODUCTION

THE HUNT BEGINS HERE

There comes a moment in every man's life when he realizes that chasing women, money, and status will never satisfy the deep hunger in his soul. I've lived that moment. I've chased the glitter. I've held what looked like treasure in my hands, only to watch it crumble into dust.

This book is for the man who's tired of the games, the heartbreak, the shallow connections—and wants something real. Something sacred. A relationship built not on lust, but on love. Not on impulse, but on purpose. Not on charm, but on Christ. I want to tell a story - not just about women, but about wisdom. A story not

> I've chased the glitter and held what looked like treasure in my hands only to watch it crumble into dust.

just about relationships, but about redemption. I want to show you what happens when a man stops chasing fool's gold and starts seeking real treasure.

In these pages,
I'll walk you through my own journey: my failures, my pride, my brokenness—and the grace of God that rebuilt me. You'll learn how to discern counterfeit from covenant, how to lead with integrity, and how to become the kind of man who can find and cherish a godly woman. But most of all, I hope you discover that the greatest treasure, you'll ever find isn't a woman— it's the man you become while walking with Christ.

Let the hunt begin.

> The greatest treasure you'll ever find isn't a woman— it's the man you become while walking with Christ.

Author Jamin Newsom

https://thetreasurehunt.life/

> THE GREATEST TREASURE
> YOU'LL EVER FIND IS
> **CHRIST.**
>
> JAMIN NEWSOME

CHAPTER 5

There's something hardwired into a man that makes him want to chase. Whether it's a dream, a job, a fight, or a woman, God placed a desire in us to pursue.

The danger isn't in the chase—it's in what we're chasing after.

I admit that for most of my life, I was chasing the wrong thing.

CHAPTER ONE

The Hunt Begins

"The treasure you're searching for won't be found in the world's direction—it's buried where only surrender can lead."

There's something hardwired into a man that makes him want to chase. Whether it's a dream, a job, a fight, or a woman, God placed a desire in us to pursue. But the danger isn't in the chase—it's in what we're chasing after. I admit that for most of my life, I was chasing the wrong thing. I thought I was after treasure. I honestly thought that if I could find the right woman—beautiful, fun, smart—then life would fall into place. That's what culture told me. That's what my own broken desires led me to believe. What I didn't know was that I was using the wrong map and following a broken compass.

See, when you don't know Christ, when you don't know your identity as a man, and when your heart hasn't been surrendered to God—you'll chase anything that glitters. You'll settle for a woman who looks like a treasure chest, only to discover she's filled with nothing but sand.

I've been there and I am here to help you.

The Bible says in Proverbs 18:22, "He who finds a wife finds what is good and receives favor from the Lord." That verse is quoted often, but most men skip the key word: finds. This word "finds" implies a search—intentional, patient, and guided.

I've seen women who looked the part. They had the smile, the body, and the charm, but over time, I learned that there was no substance. They weren't walking with God, nor were they rooted in wisdom or truth. Truth be told, I wasn't either. I was just as hollow as the relationships I found myself in. I wanted treasure, but I didn't realize I needed to become a man who could handle it first. You don't just stumble into a godly marriage by accident. You don't just "get lucky." You have to be seeking God to even recognize the right woman when she crosses your path, and yet, so many of us stop searching way to soon. We fall in love with the box before we check the contents. It's like finding a shiny chest on the beach—heavy, ornate, beautiful. But when you open it, it's filled with nothing but dirt.

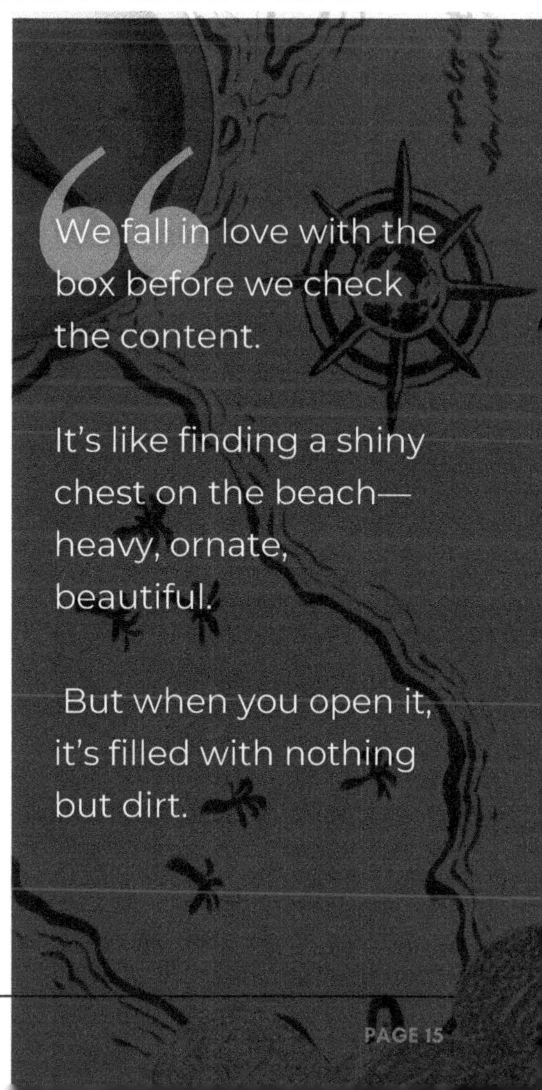

That's what happens when you chase charm without character. That's what happens when you ignore God's voice and go with your gut instead.

Let me make something clear: being attracted to the box isn't wrong. You're a man—it's natural to notice beauty; however, if beauty is all you're looking at, you'll get stuck admiring the wrapping while missing the emptiness inside.

God gave us eyes, but He also gave us discernment. If you're going to find true treasure, you need both. This book isn't about bashing women. It's not about bitterness or fear. It's about wisdom. It's about raising the standard—not just for the kind of woman you want, but the kind of man you're becoming.

The treasure hunt doesn't begin with finding her—it begins with finding Him. When you chase after God, He'll align your path. When you surrender your heart, He'll correct your compass, and when you start using the Word as your map, you'll begin to spot real treasure instead of wasting years digging in the wrong places.

Overview

Reflection Questions

Study Scripture:
He who finds a wife finds a good thing,
And obtains favor from the Lord. Proverbs 18:22

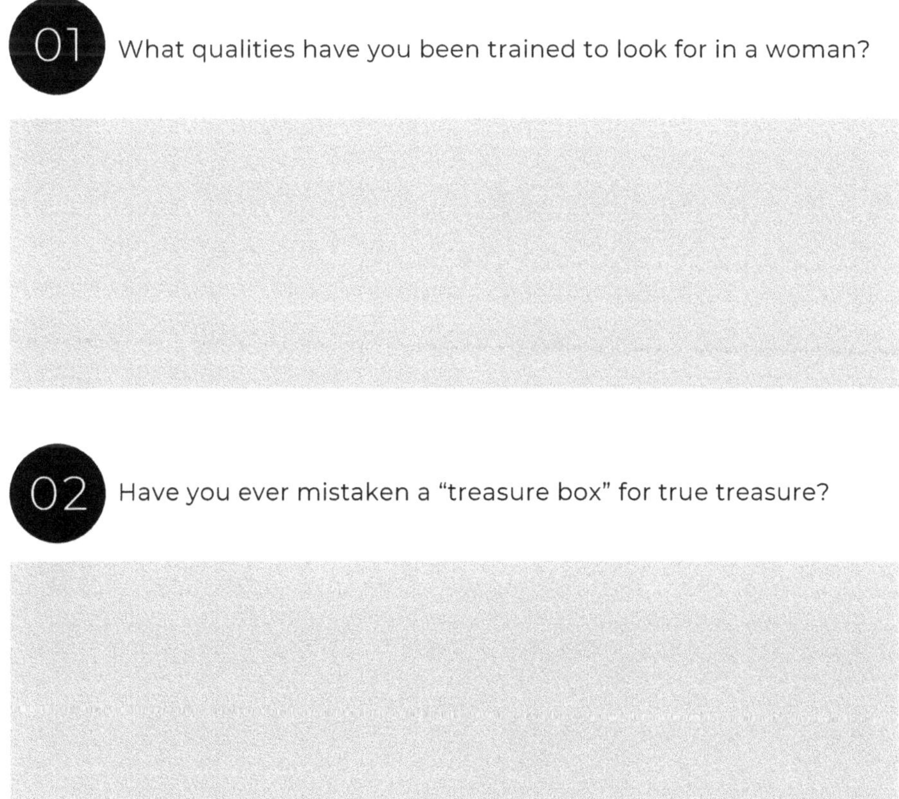

01 What qualities have you been trained to look for in a woman?

02 Have you ever mistaken a "treasure box" for true treasure?

Overview

Reflection Questions

Study Scripture:
He who finds a wife finds a good thing,
And obtains favor from the Lord. Proverbs 18:22

03 Are you using God's Word as your map in this season of searching?

Closing Charge

If you want to find real treasure, you need the real map. Don't follow your heart—follow His Word. God isn't hiding His will from you; He's written it down. The question is—will you open it?

SECTIONAL REVIEW

Share what you've learned

SECTIONAL REVIEW

Share what you've learned

> NOT EVERYTHING THAT GLITTERS IS FROM **GOD.**
>
> JAMIN NEWSOME

CHAPTER 02

Some treasure boxes sparkle so brightly that you don't realize they're empty until it's too late.

You're captivated by the outside—the beauty, the energy, the excitement—but what's inside doesn't match the packaging.

Then by the time you figure it out, your heart is already involved.

CHAPTER TWO

The False Treasure Box

"Not everything that glitters is from God—and not every woman who shines is meant to stay."

Some treasure boxes sparkle so brightly that you don't realize they're empty until it's too late. You're captivated by the outside—the beauty, the energy, the excitement—but what's inside doesn't match the packaging. Then by the time you figure it out, your heart is already involved.

I've opened more than one of those boxes.

She looked like everything I thought I wanted. She had the smile, confidence, attitude, and body. But what she didn't have was the walk with God. There was no spiritual depth, no hunger for righteousness, and no fear of the Lord. She looked like treasure, but the box was filled with sand. That's the danger of judging with your eyes instead of your spirit.

1 Samuel 16:7 says, "People look at the outward appearance, but the Lord looks at the heart." We live in a culture that trains us to chase what looks good. Social media, music, movies—they all elevate appearance over character. We're told that if she's bad, she's worth it. That if she turns heads, she's valuable. Proverbs 11:22 paints a clearer picture: "Like a gold ring in a pig's snout is a beautiful woman who shows no discretion."

In other words, beauty without godliness is just decoration on destruction. But here's the hard truth: Most of us don't learn this until we've been burned by it. I remember being in a relationship that had all the outward appeal. People thought we were goals. But behind the scenes, we were spiritually dead. There was no prayer. No purity. No purpose. Just attraction and comfort. When the storm came—and it always comes—there was nothing to hold us together because sand doesn't hold. You can build a castle on it, but it won't last. If you've ever chased a woman like that, I get it. I'm not judging you because I was you. Honestly, I wasn't even equipped to recognize a godly woman because I wasn't a godly man yet. I was attracted to drama, chaos, seduction—because those things matched the emptiness in me.

> People thought we were goals, but behind the scenes, we were spiritually dead.
>
> There was no prayer, no purity, and no purpose.
>
> Just attraction and comfort, so when the storm came, there was nothing to hold us together.

But God opened my eyes.

When He did, I had to repent—not just for chasing the wrong women, but for becoming the kind of man who was okay with it. You can't blame the box if you're the one who opened it, ignoring all the warning signs.

The problem isn't being attracted to beauty. God designed us to notice beauty. The problem is lingering too long at the wrong treasure chest. Entertaining women who don't reflect the heart of God will only delay your destiny. It doesn't mean they're bad

You weren't created to date just for fun. You weren't made to waste years in dead-end situationships. You're a man of purpose, and that means your relationships must align with your calling. If a woman's walk doesn't match your vision, then it doesn't matter how good she looks in a dress—she's not the one.

This is where discernment kicks in.

You have to ask yourself:

01 Does she love Jesus—or just post verses on Instagram?

02 Does she serve others—or is she always chasing attention?

03 Does she carry peace—or create confusion?

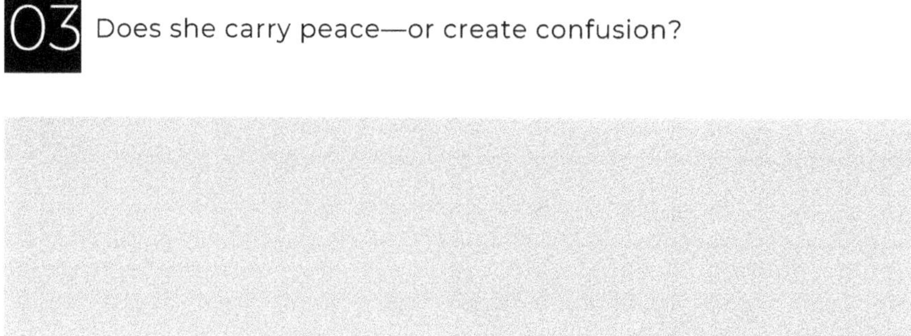

Real treasure doesn't just shine. It brings clarity, not chaos; purpose, not problems; fruit, not flattery. It's weighty.

When you see a woman who looks like a treasure chest, ask the Holy Spirit: "Is there substance inside?" Don't ignore the red flags. Don't let loneliness or lust convince you to settle. Your future is too valuable to waste digging in the wrong places.

Overview

Reflection Questions

Study Scripture:
But the Lord said to Samuel, "Do not look at his appearance or at his physical stature, because I have refused him. For the Lord does not see as man sees; for man looks at the outward appearance, but the Lord looks at the heart." 1 Samuel 16:7

01 Have you pursued someone based only on appearance?

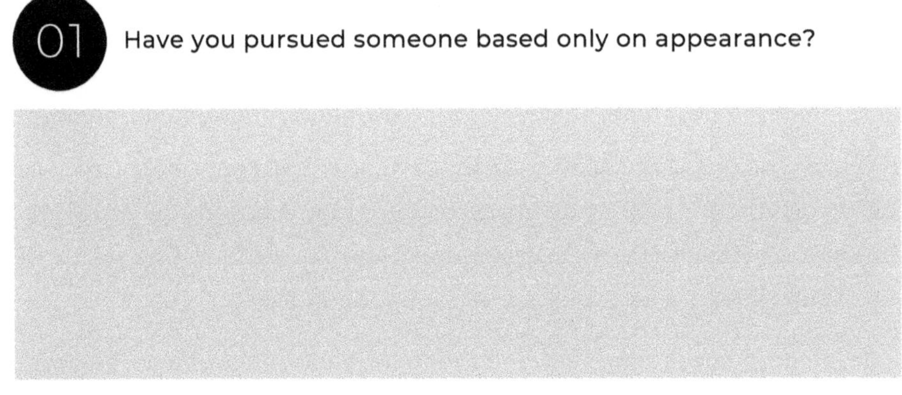

02 What patterns have I seen in the women I'm drawn to?

Overview

Reflection Questions

Study Scripture:
But the Lord said to Samuel, "Do not look at his appearance or at his physical stature, because I have refused him. For the Lord does not see as man sees; for man looks at the outward appearance, but the Lord looks at the heart." 1 Samuel 16:7

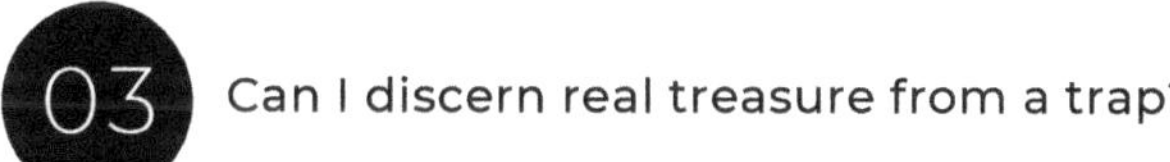

Can I discern real treasure from a trap?

Closing Charge

Not every beautiful woman is a blessing, and the same goes for boxes. Not every box is worth opening. It's important to seek the One who gives wisdom, and He will teach you how to recognize real treasure when you see it.

BEAUTY WITHOUT GODLINESS IS JUST DECORATION ON

DESTRUCTION

JAMIN NEWSOME

REVIEW

Share what you've learned

SECTIONAL REVIEW

Share what you've learned

I didn't always know I had a broken compass. In fact, I thought I was doing everything right.

I was working hard, chasing success, staying out of major trouble. I thought that made me a good man.

But the truth is—I was leading myself, and I didn't even realize I was lost.

CHAPTER THREE

My Broken Compass

"You can't follow a broken compass and expect to reach the destination."

Before God could change the women I was drawn to, He had to change the man doing the choosing. I didn't always know I had a broken compass. In fact, I thought I was doing everything right. I was working hard, chasing success, staying out of major trouble. I thought that made me a good man. But the truth is—I was leading myself, and I didn't even realize I was lost. I valued work over relationships. I valued control over vulnerability, and deep down, I was still trying to earn love and validation from everything but God.

So when it came to women, I wasn't looking for someone to build with—I was looking for someone to fill the emptiness—someone to affirm me and make me feel like a man, without actually becoming one. Here's what I've learned: you can't find true treasure when your compass is broken. You'll chase distractions and call them blessings. You'll confuse attention for affection. You'll fall for anything that looks like comfort, because deep down, you're trying to avoid confrontation with yourself.

Proverbs 14:12 says, "There is a way that seems right to a man, but its end is the way to death." That verse slapped me in the face once I came to the end of myself.

Everything I was doing seemed right at the time—grinding, achieving, dating who I wanted—but it was leading me nowhere fast. It cost me relationships, peace, even the chance to be the kind of man my wife needed me to be. I failed as a husband, not because I didn't care, but because I didn't know how to love selflessly. I was emotionally distant and spiritually passive. I was too busy working to see that my relationship was starving. I knew how to chase goals—but I didn't know how to lay myself down for another person. Without God, I didn't even know that kind of love was possible. Jeremiah 17:9 says, "The heart is deceitful above all things and desperately sick: who can understand it?" That's what happens when you lead yourself. You think your heart is guiding you toward love, but it's really guiding you toward whatever feels good in the moment.

> Everything I was doing seemed right at the time—grinding, achieving, dating who I wanted—but it was leading me nowhere fast.
>
> It cost me relationships, peace, even the chance to be the kind of man my wife needed me to be.

The truth is—I couldn't expect to lead a woman in love when I wasn't letting God lead my life.

It wasn't until I hit rock bottom that I realized the compass I was using was broken. I was building a life on my own strength, my own desires, my own wisdom—and it failed.

That failure was one of the greatest gifts God ever gave me. Because in that brokenness, I finally looked up and said, "God, I don't know what I'm doing or where I'm going. You take over."

That's when things started to change. Slowly, painfully, yet humbly, God began to rewire my perspective on everything. He showed me that my worth wasn't in what I achieved and that love meant sacrifice, not just attraction. Most importantly, He showed me that a godly relationship required a godly man—and I wasn't one yet.

I'm still learning. Still growing. But now, I'm no longer just trying to find the right woman—I'm becoming the right man.
A man who listens before he leads. A man who protects rather than manipulates.
A man who walks in purpose, not pride.

> *He broke my compass, so He could be my guide!*

You may be in that place right now—aware that something isn't right, but you are unsure how to fix it. Let me encourage you: you don't need to fix yourself. You need to surrender yourself. Let God do what only He can. Let Him show you what love looks like, by first letting Him love you.

Overview

Reflection Questions

Study Scripture:
"There is a way that seems right to a man,
but its end is the way to death." Proverbs 14:12.

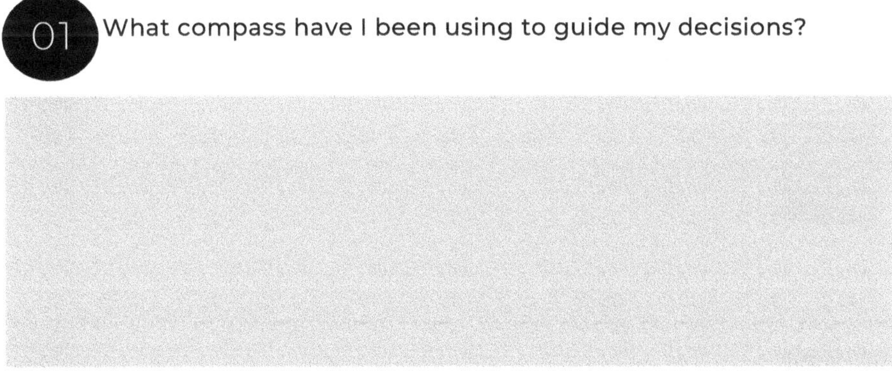

01 What compass have I been using to guide my decisions?

02 Have I acknowledged my brokenness in past relationships?

Overview

Reflection Questions

Study Scripture:
"There is a way that seems right to a man,
but its end is the way to death." Proverbs 14:12.

03 What does surrendering my love life to Jesus look like?

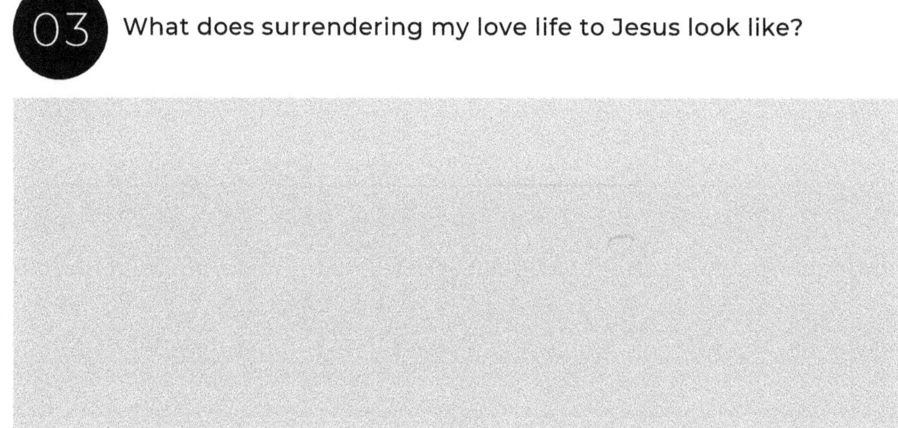

Closing Charge

Before God could show me the right woman, He had to break the compass I was using to find her. You can't find true treasure when you're still chasing fool's gold.

“NO LONGER TRYING TO FIND THE RIGHT WOMAN. I AM BECOMING THE **RIGHT MAN**

JAMIN NEWSOME

REVIEW

Share what you've learned

SECTIONAL REVIEW

Share what you've learned

CHAPTER 04

Psalm 119:105 says,
"Your word is a lamp for my feet, a light on my path."

I used to walk in the dark, trying to find love, trying to lead, trying to build a future.

But I kept stumbling because I didn't have light.

CHAPTER FOUR

The Map – God's Word

"God didn't call you to guess your way through love —He gave you a map. Open it."

After years of getting lost, I finally realized the problem wasn't just the path I was on—it was the fact that I had no real map. I was winging it. Guessing. Following my feelings and culture instead of the truth. And every time I decided without God's direction, I paid for it.

But here's what changed everything:

I opened the Word; not casually, or for a checkbox, but out of desperation. Psalm 119:105 says, "Your word is a lamp for my feet, a light on my path." I used to walk in the dark, trying to find love, trying to lead, trying to build a future, but I kept stumbling because I didn't have light. I didn't know what real love was, how to be a man of God, or what kind of woman I should be pursuing. God's Word changed that for me.

2 Timothy 3:16-17 says, "All Scripture is God-breathed and is useful for teaching, rebuking, correcting and training in righteousness..." That verse hit me hard. I needed to be taught. I needed to be corrected. And I needed to be trained in how to pursue love God's way—not the world's way.

I used to read Proverbs 31 and think, "Man, where do I find a woman like that?" What I should've been asking was, "How do I become the kind of man who can lead a woman like that?"

That question changed my life. I wanted a Proverbs 31 woman, but I hadn't even tried to become a Proverbs 1-30 man first. The truth is—until God's Word becomes your guide, you will keep ending up in the same cycles. You'll keep falling for appearances, ignoring red flags, and chasing what feels good instead of what is good.

BUT WHEN YOU LET THE WORD GUIDE YOU:

- You recognize real fruit instead of fake hype.
- You learn what a healthy pursuit looks like.
- You stop settling for women who are present but not purposeful.

God doesn't just want to bless you with a wife—He wants to prepare you for one. And that preparation comes through Scripture. When I started consistently reading the Word, I noticed I didn't just see women differently—I saw myself differently. I started to realize I wasn't ready before. I needed pruning, healing, and clarity, and the Word gave it to me. If you're serious about finding a godly woman, you need more than prayer. You need more than attraction. You need Scripture in your heart and standards in your mind. You don't need to guess anymore. God's already written the map. The question is: Will you follow it?

Overview

Reflection Questions

Study Scripture:

All Scripture is God-breathed and is useful for teaching, rebuking, correcting, and training in righteousness, 17 so that the servant of God may be thoroughly equipped for every good work. 2 Timothy 3:16-17

01 How much time do I spend reading and applying God's Word?

02 Have I submitted my view of relationships to scripture?

Overview

Reflection Questions

Study Scripture:
All Scripture is God-breathed and is useful for teaching, rebuking, correcting, and training in righteousness, 17 so that the servant of God may be thoroughly equipped for every good work. 2 Timothy 3:16-17

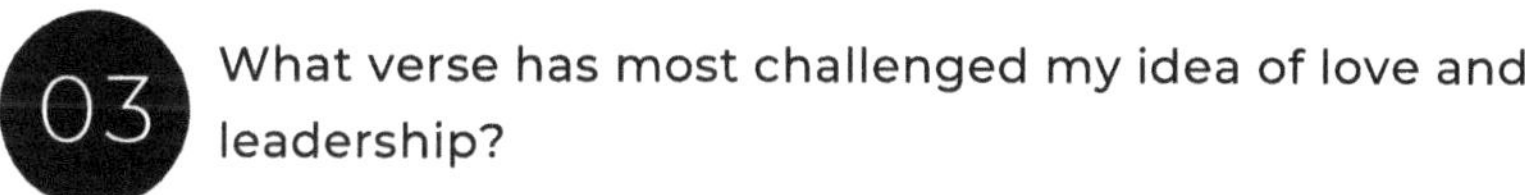

03 What verse has most challenged my idea of love and leadership?

Closing Charge

If you want to find real treasure, you need the real map. Don't follow your heart—follow His Word. God isn't hiding His will from you; He's written it down.

> I NEEDED PRUNING, HEALING, AND CLARITY, AND THE WORD GAVE IT TO ME.

MATURITY

JAMIN NEWSOME

SECTIONAL REVIEW

Share what you've learned

SECTIONAL REVIEW

CHAPTER 05

You can have a heart that genuinely wants God's best.

But if you're digging in the wrong places, you'll keep uncovering disappointment.

I learned this the hard way

CHAPTER FIVE

Digging in the Right Places

"You can't plant godly seeds in toxic soil and expect a righteous harvest."

You can have the best intentions. You can have a heart that genuinely wants God's best, but if you're digging in the wrong places, you'll keep uncovering disappointment. I learned this the hard way. Once I started getting serious about my faith—after God began breaking and rebuilding me—I thought, "Okay, I'm ready. Bring on the godly woman."

Unfortunately, I was still trying to find her in the same places I used to find distractions. That's not how it works.

You don't look for a diamond in the garbage. You don't find gold at the bottom of a muddy ditch. You especially don't find a godly woman in environments that don't honor God.

The Bible says in Psalm 1:1-2, "Blessed is the one who does not walk in step with the wicked or stand in the way that sinners take or sit in the company of mockers, but whose delight is in the law of the Lord..." Translation? Where you walk matters. The company you keep, the places you go, the standards you uphold—all of it affects the kind of people you attract.

I'm not saying God can't move anywhere. He can, but more often than not, when you look in environments that prioritize lust over love, attention over character, and popularity over purpose, you're going to find women who reflect those values.

Proverbs 13:20 backs this up: "Walk with the wise and become wise, for a companion of fools suffers harm." It's not just about being in church or not being in a club. It's about the spirit of the space you're standing in. Does the environment you're in cultivate growth, faith, truth, or just attention and ego?

Here's the thing: Sometimes, you'll stumble across a woman who seems "almost right." She's got a few spiritual qualities. She's kind and talks about God here and there. But deep down, something feels off. However, instead of walking away, you linger. You hope. You try to shape the relationship into something real.

Let me say this as clearly and as simply as I can: Do NOT linger at the wrong treasure chest out of fear that you will not find another one. That's not faith—that's desperation. God doesn't bless fear-driven decisions. I've been there, though—staying too long in something God never told me to pursue. I justified the red flags because I was tired of waiting.

Trying to force something that looked close enough to what I wanted. God isn't calling you to settle for "close enough." He's calling you to His best. Sometimes, the problem isn't that God hasn't sent the right woman.

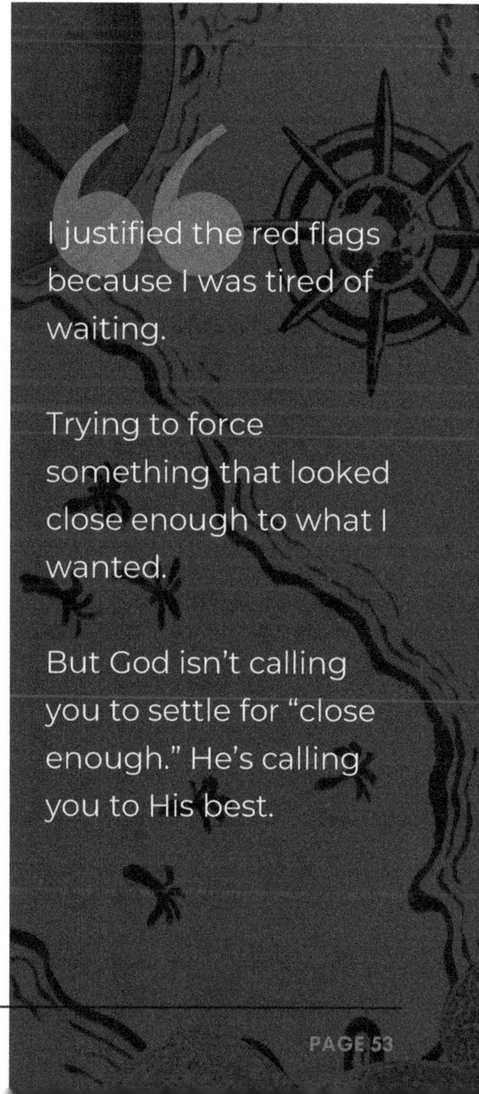

CALL TO ACTION

It's that we're still digging in places where she would never be.
Real treasure is often buried in places the world overlooks. A woman of God may not have the largest following. She might not be loud or flashy, but she's rooted. She's steady. She's about her Father's business, and you'll never find her if you're still entertaining chaos. So stop showing up in places that contradict what you're praying for. If you want a woman of purity, stop hanging around spaces driven by lust. woman of prayer, stop pursuing women who ghost God all week. If you want a woman of wisdom, stop entertaining women who treat life like a game.

And maybe the deeper question is this:

Are you becoming the kind of man who even knows where to dig?

Because the search isn't just about finding her—it's about growing into who you are called to be along the way.

A GODLY WOMAN IS

ROOTED IN GOD

FAITHFUL TO GOD

STEADY IN GOD

Overview

Reflection Questions

Study Scripture:
But the Lord said to Samuel, "Do not look at his appearance or at his physical stature, because I have refused him. For the Lord does not see as man sees; for man looks at the outward appearance, but the Lord looks at the heart." 1 Samuel 16:7

01 Where have I been looking for relationships?

02 Where should I be looking for a godly woman?

Overview

Reflection Questions

Study Scripture:
But the Lord said to Samuel, "Do not look at his appearance or at his physical stature, because I have refused him. For the Lord does not see as man sees; for man looks at the outward appearance, but the Lord looks at the heart." 1 Samuel 16:7

03 Am I growing into the kind of man who can recognize and cherish a godly woman?

Closing Charge

Godly women don't grow in toxic soil. If you're tired of counterfeit treasure, change where you dig. God is not hiding the right woman from you—He's preparing both of you for the right time.

> YOU ESPECIALLY DON'T FIND A GODLY WOMAN IN ENVIRONMENTS THAT DON'T **HONOR GOD**
>
> JAMIN NEWSOME

SECTIONAL REVIEW

Share what you've learned

SECTIONAL REVIEW

Share what you've learned

CHAPTER 06

You can't turn sand into gold.

You can't force treasure into a box that God never designed to hold it.

But that's exactly what many of us try to do. We meet a woman who has "potential," and instead of walking away in wisdom, we stick around trying to build her into something she's not.

CHAPTER SIX

Treasures Are Found, Not Fabricated

"You weren't called to build a wife—you were called to recognize the one God already formed."

You can't turn sand into gold. You can't force treasure into a box that God never designed to hold it. But that's exactly what many of us try to do. We meet a woman who has "potential," and instead of walking away in wisdom, we stick around trying to build her into something she's not.

We become pastors, fathers, counselors, spiritual coaches—all while calling it love, but it's not love. It's pride. It's fear and It's control.

Let me say it clearly:

You were never called to fabricate treasure.

Proverbs 19:14 says, "Houses and wealth are inherited from parents, but a prudent wife is from the Lord." That means a godly woman isn't something you create—she's someone you discover, someone God leads you to. She already has a walk with Him. She already fears the Lord. She already knows who she is in Christ. Your job is not to shape her. Your job is to recognize her.

Now, hear me: this doesn't mean she'll be perfect. None of us are, but there's a difference between someone growing and someone needing to be remade entirely. If the woman you're dating has no spiritual fruit, no prayer life, no desire for God, but you're staying because you hope she'll "get there someday"—you're not loving her. You're using her.

You're using her to fill a void, while convincing yourself that you're doing the godly thing by being "patient." You're not her Savior. You're not the Holy Spirit. If she only grows because of your leadership, what happens when you're tired? Or discouraged? Or broken? The weight of her faith walk should not rest on your shoulders. That is not your burden to bear. 2 Corinthians 6:14 warns us, "Do not be unequally yoked with unbelievers..." But this applies beyond salvation status. Are you equally yoked in vision? In spiritual discipline? In purpose?

Because if you're dragging her spiritually while she drags you emotionally, what you have isn't a partnership—it's a project, and projects drain.

I've been there. I stayed in relationships way too long, trying to fix, correct, and shape someone into a wife. I confused brokenness with destiny. I saw glimmers of potential and let it blind me to the reality of where she was. Every time, it left me exhausted.

Genesis 2:22 gives us a better model: "Then the Lord God made a woman from the rib he had taken out of the man, and He brought her to the man." Did you catch that? God made her, and then he brought her. Adam didn't build her. He didn't mold her.

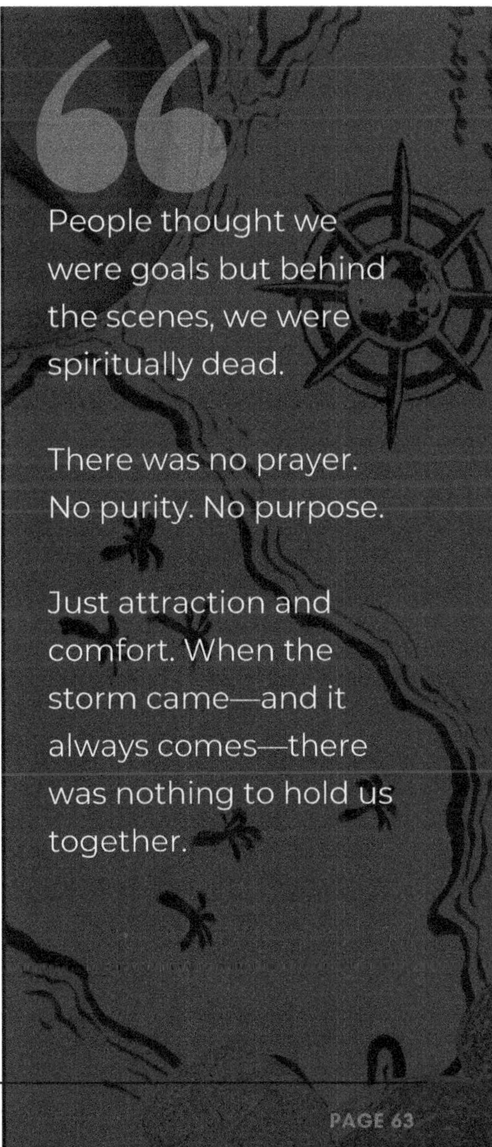

He received what God had already prepared. You don't need to fabricate a wife. You need to trust God to bring you one who's been formed by His hand, not your hustle. If that means waiting, then wait. Don't compromise your calling by trying to build what God never asked you to. Don't invest time in someone God never permitted you to pursue, and don't spiritualize disobedience by calling it patience.

WHEN SHE'S FROM GOD

- When the woman is from God, ·you won't have to force the foundation.
- When the woman is from God, · you won't have to drag her to prayer.
- When the woman is from God, you won't have to convince her to honor purity.
- When the woman is from God, you won't have to pretend.

She won't be perfect, but she will be prepared.
Stop fabricating. Start discerning.

Overview

Reflection Questions

Study Scripture:
"Then the Lord God made a woman from the rib he had taken out of the man, and he brought her to the man."Genesis 2:22

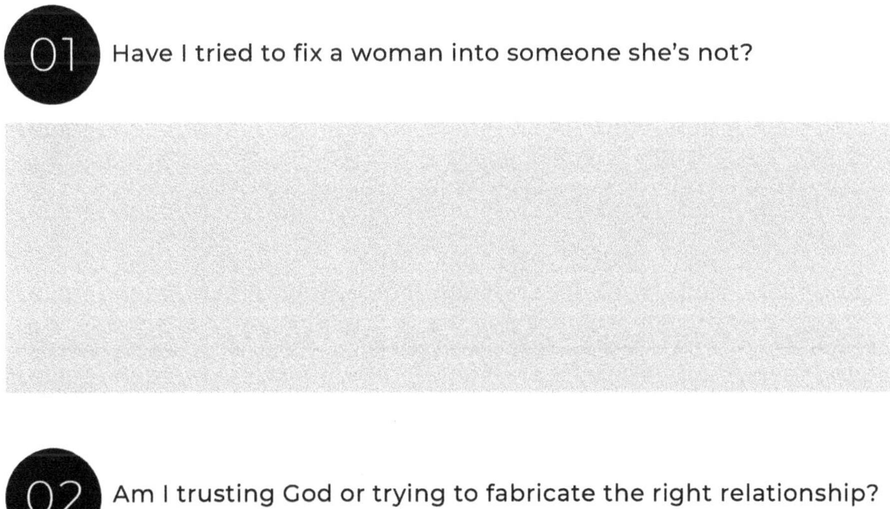

01 Have I tried to fix a woman into someone she's not?

02 Am I trusting God or trying to fabricate the right relationship?

Overview

Reflection Questions

Study Scripture:
"Then the Lord God made a woman from the rib he had taken out of the man, and he brought her to the man."Genesis 2:22

03 Can I recognize the evidence of godly character in a woman?

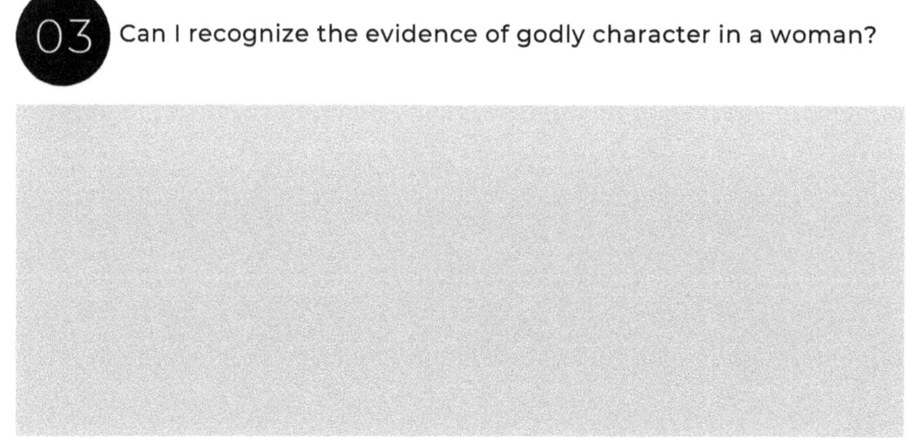

Closing Charge

Stop exhausting yourself trying to mold a woman into something she's not. When the treasure is real—you won't have to force it.

"THERE'S A DIFFERENCE BETWEEN SOMEONE GROWING AND SOMEONE NEEDING TO BE REMADE ENTIRELY.

GROWTH

JAMIN NEWSOME

SECTIONAL REVIEW

Share what you've learned

CHAPTER 07

It looks valuable. It looks real. But when you hold it long enough, when life applies pressure, when the fire hits—it reveals itself for what it is:

Empty, cheap, and fake.

What I was calling love wasn't love. It was lust. It was validation.

It was fear of being alone.

CHAPTER SEVEN

From Fool's Gold to Grace

"Grace isn't just God's way of covering your past—it's His invitation to start again."

I thought I found treasure once. A few times, actually.

She looked the part. Said all of the right things. She even acted how I wanted and for a moment, it felt like I finally got it right. But it wasn't long before I realized I'd been fooled—again.

Because fool's gold shines too.

It looks valuable. It seems real until you hold it long enough, when life applies pressure, when the fire hits—it reveals itself for what it really is: empty, cheap, and fake. What I was calling love wasn't love. It was lust. It was validation. It was fear of being alone, and worse, it was all being filtered through a broken version of me. I chased relationships without being healed. I pursued women without having a vision, and I stayed in situations that God never ordained—because I was more afraid of being single than being disobedient.

Isaiah 61:3 speaks of God's ability to give "beauty for ashes." But before He gave beauty, He let me see the ashes—everything I burned down with my own hands. The relationships were ruined. The trust was broken. The time wasted. I found grace.

Romans 5:8 says, "While we were still sinners, Christ died for us." That hit differently when I was in the middle of my mess. He didn't wait for me to get right. He didn't wait for me to finally stop chasing women. He loved me while I was still broken, still selfish, still prideful. Grace didn't just forgive me —It transformed me. It gave me the strength to walk away from situations that once controlled me. It gave me peace in seasons of singleness and wisdom to see through fake treasure. Most of all, it gave me identity. 2 Corinthians 5:17 says, "If anyone is in Christ, he is a new creation. The old has passed away; behold, the new has come." That's not poetry. That's a promise.

> *I didn't need to become a better version of my old self. I needed to become a new man.*

I didn't need to become a better version of my old self. I needed to become a new man. A man rooted in truth. A man who didn't chase to be filled, but was already full in Christ. A man who didn't depend on women to affirm him, but relied on God to define and refine him. Maybe that's where you are right now. Maybe you've been chasing fool's gold and carrying shame from relationships you know you weren't supposed to be in. Maybe you feel like you've wasted too much time or hurt too many people to be restored.

Let me tell you something about grace, it covers all. God doesn't just tolerate you—He chooses you. He doesn't want to just fix your past—He wants to give you a future. One filled with purpose, purity, and real love. You don't have to be the man you used to be. Let Him make you new.

Overview

Reflection Questions

Study Scripture:
"While we were still sinners, Christ died for us." Romans 5:8

01 Have I been chasing fool's gold?

02 Where have I experienced God's grace in my story?

Overview

Reflection Questions

Study Scripture:
"While we were still sinners, Christ died for us." Romans 5:8

03 Am I letting God rebuild me into something new?

Closing Charge

Fool's gold may have fooled you before, but grace is greater than your past. Let God rebuild you into the kind of man who stops chasing illusions and starts walking in truth.

"I DIDN'T NEED TO BECOME A BETTER VERSION OF MY OLD SELF. I NEEDED TO BECOME A **NEW MAN**

JAMIN NEWSOME

SECTIONAL REVIEW

Share what you've learned

SECTIONAL REVIEW

Share what you've learned

CHAPTER 08

You prayed. You waited. You grew. And now—maybe God brings her into your life. A woman of character. Of prayer. Of peace. A real one. You'd think the hard part is over, right?

But what if I told you that finding the treasure isn't the finish line—it's just the beginning?

Because a real man doesn't just receive a godly woman. He protects her.

CHAPTER EIGHT

Guarding the Treasure

"Real love doesn't take what it wants—it protects what it values."

It's one thing to find treasure. It's another thing to guard it.

You prayed. You waited. You grew and now, maybe God brings her into your life. A woman of character, of prayer and peace. A real one. You'd think the hard part is over, right? But what if I told you that finding the treasure isn't the finish line—it's just the beginning. A real man doesn't just receive a godly woman. He protects her.

> 1 Thessalonians 4:3-4 says, "For this is the will of God, your sanctification: that you abstain from sexual immorality; that each one of you know how to control his own body in holiness and honor." God doesn't just want you to pursue a woman—He wants you to pursue purity. And not just her purity—yours too.

Let's be real. Once emotions are high, feelings are mutual, and chemistry is present, purity doesn't just "happen." It's fought for. It's guarded. And in a world that says, "if it feels right, do it," you've got to stand firm in what's actually righteous.

Guarding the treasure means setting boundaries:

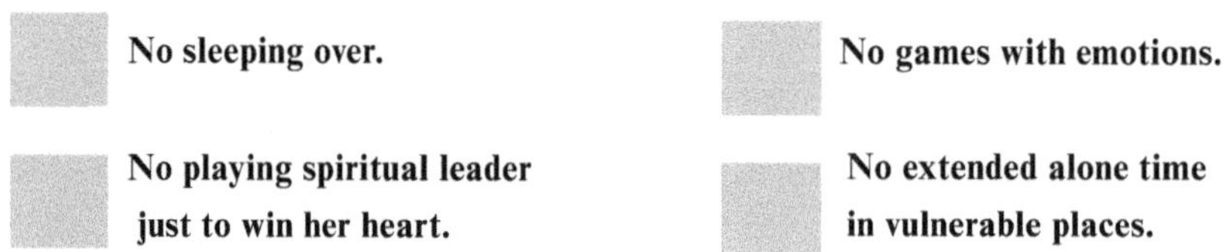

- **No sleeping over.**
- **No games with emotions.**
- **No playing spiritual leader just to win her heart.**
- **No extended alone time in vulnerable places.**

Purity isn't about restriction—it's about respect. And more than that, it's about preparation. When you guard her purity, you're preparing to guard her heart, her name, and her legacy. Also, guarding her purity before marriage is a testimony to your fidelity and commitment during marriage. Proverbs 4:23 says, "Above all else, guard your heart, for everything you do flows from it." That goes for both of you. Don't lead her with emotion—lead her with vision. If God has shown you she's the one, then show her that He's still the One you love most. I began dating differently when I finally understood what it meant to lead with integrity—I stopped asking, "How far can we go?" and instead started asking, "How holy can we be?" And let me tell you! There is nothing more attractive to a woman of God than a man who knows how to lead in love without needing to touch her body to do it.

Don't lead her with emotion—lead her with vision.

If God has shown you she's the one, then show her that He's still the One you love most.

I began dating differently when I finally understood what it meant to lead with integrity.

Ephesians 5:25 sets the standard: "Husbands, love your wives, just as Christ loved the church and gave Himself up for her…" That's sacrificial love. Protective love. And while you may not be her husband yet, your actions should still reflect the heart of one.

And if you mess up? Don't fake it. Repent. Take ownership. Invite accountability. Grace doesn't mean you keep playing with fire.

Overview

Reflection Questions

Study Scripture:
"Husbands, love your wives, just as Christ loved the Church and gave Himself up for her..." Ephesians 5:25

01 Am I leading the woman in my life toward purity or temptation?

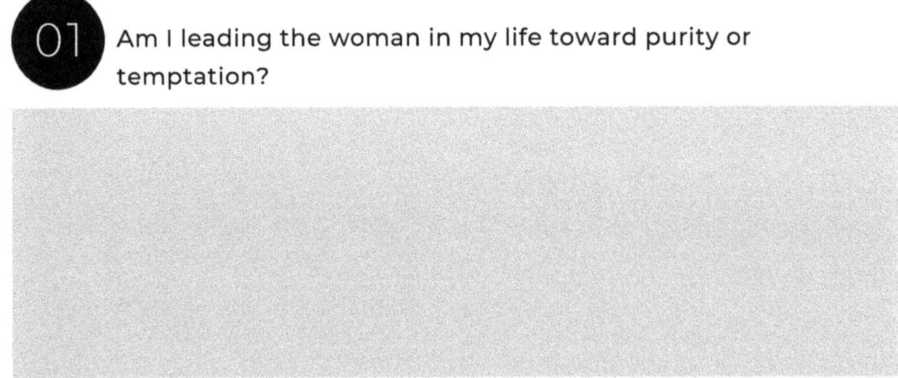

02 What boundaries do we need to honor God?

Overview

Reflection Questions

Study Scripture:
"Husbands, love your wives, just as Christ loved the church and gave Himself up for her..." Ephesians 5:25

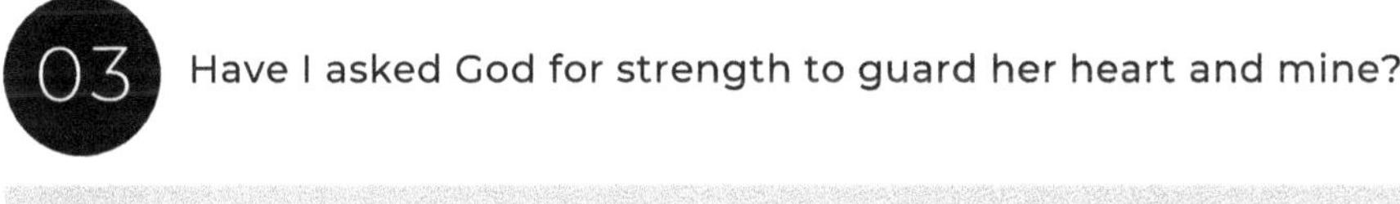

03 Have I asked God for strength to guard her heart and mine?

Closing Charge

If you truly value her—you'll protect her, not just pursue her. Treasure is always under attack. Guard it with everything you've got.

“PURITY ISN'T ABOUT RESTRICTION—IT'S ABOUT RESPECT, AND

PREPARATION.

JAMIN NEWSOME

SECTIONAL REVIEW

Share what you've learned

SECTIONAL REVIEW

Share what you've learned

When God gives you a woman, He's not just giving you someone to enjoy—He's giving you someone to protect, to pour into, and to build with.

Marriage is the place where two people who were running toward God suddenly realize—they're running side-by-side.

Now, their race becomes one.

CHAPTER NINE

THE TREASURE SHARED

"Marriage isn't where the treasure hunt ends—it's where the legacy begins."

You spend the time. You endured the process. You let God refine you, and now, He's brought someone into your life that reflects His goodness, His grace, and His design. But here's what most men don't understand:

The treasure you found is not just for you—it's meant to be shared.

Ecclesiastes 4:9-12 says, "Two are better than one, because they have a good return for their labor... A cord of three strands is not quickly broken." This is marriage. Not just romance, but partnership. Not just feelings, but foundation. It's you, her, and God—woven together for kingdom purpose.

She's not just your wife—she's your mission partner. She doesn't just meet your emotional needs—she's called to help multiply your purpose. And you're called to multiply hers, too. The world will tell you that marriage is about getting what you want. About comfort. About benefits. But Ephesians 5:25 says otherwise: "Husbands, love your wives, just as Christ loved the church and gave Himself up for her..." That's not selfish love. That's sacrificial love. That's leadership rooted in service. When God gives you a woman, He's not just giving you someone to enjoy—He's giving you someone to protect, to pour into, and to build with. Marriage is the place where two people who were running toward God suddenly realize—they're running side-by-side. Now, their race becomes one.

BUILDING HER MEANS

PROTECT, POUR, AND BUILD WITH MEMORIZATION

She's not just your wife—she's your mission partner. She doesn't just meet your emotional needs—she's called to help multiply your purpose. You're called to multiply hers too.

LET'S TAL K ... LEGACY

He seeks Godly offspring

Malachi 2:15 says, "Has not the one God made you? You belong to Him in body and spirit. And what does he seek? Godly offspring." That's not just about children. That's about spiritual impact. About legacy. About planting seeds—truth, faith, righteousness—that will outlive you.

Your love isn't just about romance.

It's about reflecting Christ. It's about creating a safe space where truth, purpose, and peace can thrive. It's about modeling what covenant really looks like in a world full of counterfeits. And no—marriage won't be perfect.

Build. Dream. Sacrifice.
Pray. Serve. and Keep Christ at the center.

There will be arguments, miscommunication, tears, and frustration, but when Christ is the foundation, the storm may shake the house, but it won't break it. The treasure isn't just the woman; it's what God does through the two of you when you walk in unity.

So build together, dream together, and sacrifice for one another.
Be sure to pray together, serve one another, and keep Christ at the center. Let the world watch what happens when a man of God and a woman of God decide to share the treasure instead of hoarding it.

Overview

Reflection Questions

Study Scripture:
Has not the one God made you? You belong to Him in body and spirit and what does he seek? Godly offspring." Malachi 2:15

01 Do I see marriage as the beginning of shared purpose, or the end of a pursuit?

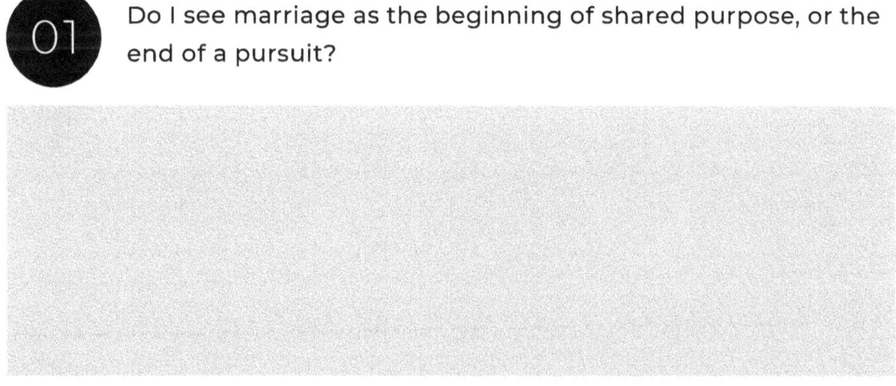

02 Am I prepared to love sacrificially, serve faithfully, and lead spiritually?

Overview

Reflection Questions

Study Scripture:
Has not the one God made you? You belong to Him in body and spirit, and what does he seek? Godly offspring." Malachi 2:15

What kind of legacy do I want to build with the woman God gives me?

Closing Charge

The real treasure isn't just the woman—it's what God will do through the two of you together. Marriage is the soil for legacy, mission, and transformation.

"THE TREASURE YOU FOUND IS NOT JUST FOR YOU. IT'S MEANT TO BE **SHARED.**

JAMIN NEWSOME

SECTIONAL REVIEW

Share what you've learned

SECTIONAL REVIEW

Share what you've learned

CHAPTER 10

I used to think finding the right woman would make me whole, but it wasn't until I let God make me whole first that I was finally ready for her.

The pursuit of God redefined everything:

I didn't just become a better man—I became a new man.

CHAPTER TEN

The Real Reward

"The greatest reward of the search was never her—it was who God made you along the way."

You started this journey looking for treasure. Maybe that treasure was a woman. Maybe it was healing. Maybe it was clarity. But somewhere along the way, God revealed something deeper: you weren't just looking for treasure—you were being shaped into it.

Yes, a godly wife is a gift. Yes, marriage is a blessing.

But if we're honest, many of us started this hunt thinking she would be the reward. Once we found her, everything would fall into place. But here's what I've learned: The greatest treasure in life isn't found in her arms—it's found at His feet.

Matthew 6:33 says, "Seek first the kingdom of God and His righteousness, and all these things will be added to you." We flip that all the time. We seek things. The wife. The home. The sex. The peace. And then, if there's time, we seek God. But that is just plain backwards. And it's empty.

Because the real reward isn't the blessing—it's the One who gives it.

I used to think finding the right woman would make me whole. But it wasn't until I let God make me whole first that I was finally ready for her. The pursuit of God redefined everything: I didn't just become a better man—I became a new man. I didn't just want a godly woman—I wanted to walk in godliness myself. I didn't just want love—I wanted to love like Christ. Philippians 3:8 says, "I count everything as loss because of the surpassing worth of knowing Christ Jesus my Lord." And I feel that now. Everything I used to chase—women, validation, status—none of it compares to knowing Him. That's not a church cliché. That's real. When you know Jesus, and I mean really know Him:

WHEN YOU KNOW JESUS

- You stop chasing people to fill you.
- You stop performing to be accepted.
- You stop compromising your standards to feel loved.

Because your heart is already full. So yes, God may bring you a woman. A good one. A Proverbs 31, purpose-driven, Holy Spirit-filled woman. But she will not be your reward. She will not be your savior. She will not complete you. She will compliment you. Your real reward is Christ. The reward is becoming the kind of man who reflects His love. The reward is walking in purity, power, and purpose. The reward is building a life that outlives you. And when the woman shows up? She won't just be someone you need—she will be someone you can bless. So take this to heart: You are not just a man looking for treasure. You are a man becoming a treasure. A man who chose surrender over pride. Obedience over shortcuts. Purpose over pleasure. That is the kind of man God can trust with a wife, a calling, and a legacy.

This is the reward: walking with Jesus. Everything else? That's just overflow.

Overview

Reflection Questions

Study Scripture:
I count everything as loss because of the surpassing worth of knowing Christ Jesus my Lord. Philippians 3:8

01 Have I been seeking the gift more than the Giver?

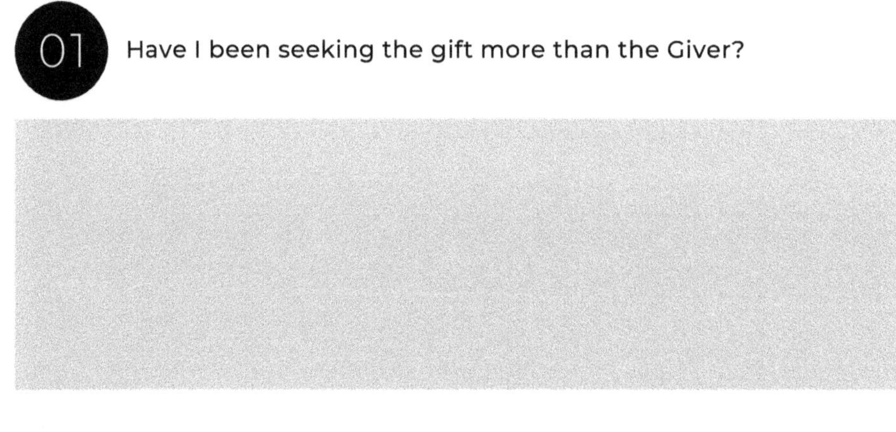

02 What has God taught me about myself during this journey?

Overview

Reflection Questions

Study Scripture:
I count everything as loss because of the surpassing worth of knowing Christ Jesus my Lord. Philippians 3:8

03 How can I continue to grow into the man God is calling me to be?

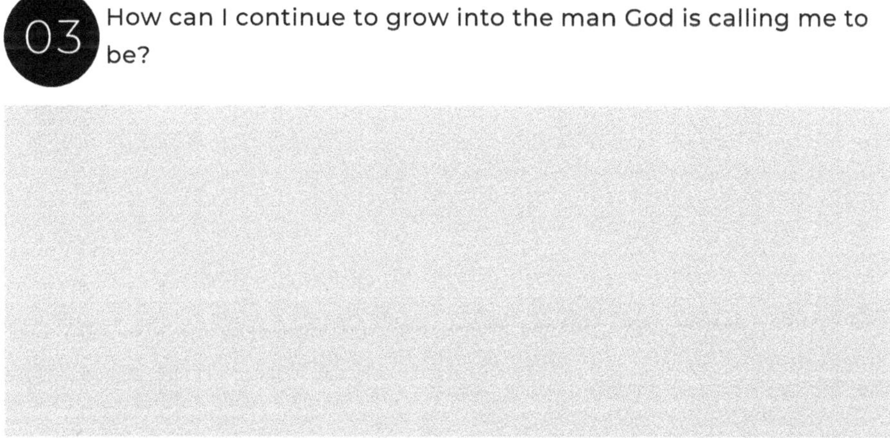

Closing Charge

You weren't just looking for treasure—you were being shaped into it. The reward isn't just who you find, but who you become. When you walk with God, the treasure is never far behind.

> YOUR REAL REWARD IS CHRIST. THE REWARD IS BECOMING THE KIND OF MAN WHO REFLECTS HIS LOVE

BE IN CHRIST.

JAMIN NEWSOME

SECTIONAL REVIEW

Share what you've learned

SECTIONAL REVIEW

Share what you've learned

CONCLUSION

If you've made it to this point, I want to say something clearly: You are not the same man you were when you started. You've wrestled with truth. You've been challenged to let go of false ideas about love, women, and even yourself. You've been invited to surrender your compass, pick up the map, and dig where God leads.

Whether you're single, dating, or even married—the journey has never really been about the woman. It's always been about the man—the kind of man you are becoming. The kind who walks in wisdom. The kind who doesn't settle for what sparkles but searches for what's sacred.

The kind who doesn't just chase a wife, but who seeks the heart of God first, and lets everything else flow from there. The kind who walks in wisdom. The kind who doesn't settle for what sparkles but searches for what's sacred. The kind who doesn't just chase a wife, but who seeks the heart of God first, and lets everything else flow from there.

So as you close this book, don't stop digging.

- Dig deeper into scripture.
- Dig deeper into prayer.
- Dig deeper into who God is calling you to be.

Because when you seek Him first, everything else—including the treasure—will be added in time.

ABOUT THE AUTHOR

JAMIN NEWSOME

Jamin Newsome is a U.S. military veteran, private security professional, and the founder of Millstone Security—a faith-driven company committed to protecting lives and fighting human trafficking. Drawing from his own journey through divorce, brokenness, and redemption, Jamin writes with boldness, authenticity, and a passion for helping men discover what truly matters. After years of chasing status, success, and superficial relationships, Jamin came face-to-face with the truth: real love requires real transformation.

From that moment forward, he committed to becoming the man God designed him to be, and helping others do the same. In Treasure Hunt: Finding Love God's Way, Jamin brings men into the trenches of his own story, offering a road map built on Scripture, tested in failure, and grounded in grace. Whether speaking, writing, or leading, his message is the same: God's way is worth it, and manhood begins at the cross. When he's not writing, training, or working in executive protection, Jamin is a proud father, a devoted believer, and a man in pursuit of legacy.

"YOU HAVE SUCH A POWERFUL STORY WITHIN YOU FROM GOD"

LIVING WATER BOOKS

OFFICE NUMBER
501-488-0031
Livingwaterbooks.org
Livingwaters@livingwaterbooks.org

PUBLISHING CO